RADIO COLLECTING:

A Beginner's Guide

SREEKUMAR V T

PREFACE:

Welcome to "Radio Collecting: A Beginner's Guide." In this book, we aim to introduce you to the exciting world of vintage radios and provide you with the tools and knowledge necessary to start your own collection.

Radios have played a significant role in our lives for over a century, providing entertainment, news, and a connection to the world beyond our immediate surroundings. As technology continues to advance, the world of radio collecting is becoming increasingly exciting. Vintage radios can be restored to their former glory, and even enhanced with modern features, providing a unique and rewarding hobby for collectors and enthusiasts.

This book is designed to be a comprehensive guide for anyone who is interested in starting their own radio collection. We cover everything from the history of radios to the basics of electronics, from restoring and repairing vintage radios to finding resources and connecting with other collectors.

Our hope is that this book will inspire you to dive deeper into the world of radio collecting and provide you with the knowledge and skills necessary to build your own collection. Whether you're a seasoned collector or just starting out, we believe that this book will be a valuable resource on your journey.

Thank you for choosing "Radio Collecting: A Beginner's Guide." We hope that you enjoy reading it as much as we have enjoyed creating it, and that it provides you with a solid foundation for your future endeavors in the world of radio collecting.

DISCLAIMER:

The information contained within this book is intended to provide general guidance and advice on the topic of radio collecting. While the authors have made every effort to ensure the accuracy of the information provided, they cannot be held responsible for any errors, omissions or inaccuracies. The content in this book is not a substitute for professional advice, and the authors make no representations or warranties of any kind, express or implied, about the completeness, accuracy, reliability, suitability or availability with respect to the information contained within.

The reader assumes full responsibility for any actions taken based on the information presented in this book. The authors will not be liable for any losses, injuries or damages resulting from the use or misuse of the information provided. The reader should always exercise caution and use their own judgment when restoring or repairing vintage radios, and seek professional advice when needed.

The authors do not endorse or recommend any specific brands or products mentioned in this book, and any reference to such products is for informational purposes only. The authors have no financial or commercial interest in any of the products or services mentioned in this book.

Finally, the authors reserve the right to update and revise the information contained within this book at any time without notice.

CONTENT

1. INTRODUCTION:

Why Collect Vintage Radios?

T he world of vintage radio collecting is a fascinating and rewarding hobby that has attracted enthusiasts for decades. Vintage radios can be more than just objects of nostalgia; they can also be a gateway to learning about the history and technology of radio and electronics.

There are many reasons why people collect vintage radios. For some, it's the thrill of the hunt, the excitement of discovering rare or unusual models at flea markets, garage sales, or online auctions. For others, it's the joy of restoring and repairing these old radios, bringing them back to their former glory and keeping them in working condition.

Some collectors are drawn to the aesthetics of vintage radios, with their sleek designs and intricate details, while others are interested in the social and cultural history behind these objects, and the role that radio played in shaping society and communication.

Whatever your motivation, radio collecting is a hobby that can be enjoyed by people of all ages and backgrounds. It requires no special skills or expertise, just a curiosity and passion for the subject. And with a wide variety of vintage radios available, from simple table-top models to ornate console units, there's something for every collector to appreciate.

In this book, we will explore the world of vintage radio collecting in depth, covering everything from identifying and acquiring vintage radios, to restoring and repairing them, to displaying and enjoying your collection. We will also look at the history and technology of radio, and how it has evolved over time.

Whether you are a seasoned collector or just starting out, this book will provide you with a wealth of information and resources to help you pursue your hobby and build a collection of vintage radios that you can be proud of.

Furthermore, collecting vintage radios can also be a great way to connect with others who share your passion for radio and electronics. There are many online forums, social media groups, and local clubs dedicated to vintage radio collecting, where you can share your knowledge and experiences, learn from others, and make new friends.

Vintage radios also offer a glimpse into the history of technology and innovation. They represent a time when radio was the primary means of mass communication and entertainment, and they provide a window into the social, cultural, and economic forces that shaped the world in which they were created.

In addition, vintage radios can be a valuable investment for collectors. While prices for individual radios can vary widely depending on their rarity, condition, and other factors, there is a growing market for vintage radios among collectors and enthusiasts. As with any investment, it's important to do your research and make informed decisions, but owning a valuable and historic piece of technology can be a source of pride and satisfaction for collectors.

Overall, there are many compelling reasons to collect vintage radios, from the thrill of the hunt to the historical and cultural significance of these objects. With this book as your guide, you can begin your journey into the world of vintage radio collecting and discover the joys and rewards of this fascinating hobby.

2. GETTING STARTED:

What You Need to Know
Before You Begin

Before diving headfirst into the world of vintage radio collecting, there are a few things you should consider and prepare for. In this chapter, we will cover some important steps and tips to help you get started on the right foot.

The first thing you should do is research. Learn as much as you can about vintage radios and their history. This will help you identify and appreciate the different types and models of radios you encounter, as well as help you determine what kind of radio you want to collect. Look for books, magazines, and online resources dedicated to vintage radio collecting, and read up on the subject.

Once you have a good understanding of vintage radios, it's time to set a budget. Decide how much you are willing to spend on your collection, and be realistic. Vintage radios can vary widely in price, from a few dollars for a simple table-top model to thousands of dollars for a rare or highly sought-after radio. Setting a budget will help you stay focused and avoid overspending.

Next, determine the space you have available for your collection. Vintage radios can take up a lot of space, so it's important to plan accordingly. Consider factors such as display options, storage needs, and how much space you have to work with.

Another important step is to identify reputable sellers and sources for vintage radios. Look for dealers and collectors who have a good reputation and a history of fair dealing. Avoid buying from unknown sellers or those who do not have a proven track record.

Finally, invest in the right tools and supplies for your collection. Basic tools such as screwdrivers, pliers, and wire cutters will come in handy when restoring and repairing radios. You may also want to invest in specialized tools such as a tube tester or a signal generator, depending on the level of restoration work you plan to do.

By taking these steps and preparing yourself before you begin collecting vintage radios, you will be better equipped to enjoy and appreciate your collection to its fullest potential.

3. IDENTIFYING VINTAGE RADIOS:

Understanding Different Types and Models

One of the most important skills you'll need as a vintage radio collector is the ability to identify different types and models of radios. This chapter will provide you with an overview of the different types of vintage radios and help you understand what to look for when identifying them.

The first type of vintage radio you're likely to encounter is the table-top radio. These radios are relatively small and designed to sit on a table or shelf. They were popular from the 1920s through the 1950s and come in a variety of shapes and styles. Some of the most popular designs include the cathedral radio, tombstone radio, and the Art Deco radio.

Another type of vintage radio is the console radio. These radios are larger and were designed to be the centerpiece of a living room or other gathering space. They typically feature a combination of radio and record player, and some even have a built-in phonograph. Console radios were popular from the 1930s through the 1950s.

There are also portable radios, which were designed to be carried around and used on the go. These radios were popular from the 1940s through the 1970s and come in a variety of styles and sizes. Some are battery-powered, while others require an external power source.

In addition to understanding the different types of vintage radios, it's also important to be able to identify specific models. This can be challenging, as many radios were produced in limited numbers and may be difficult to find today. However, there are a few key things to look for when identifying a vintage radio.

First, look for a manufacturer's label or nameplate. This will typically be located on the back of the radio and will provide important information about the manufacturer, model, and year of production. You can use this information to research the radio and learn more about its history and value.

Another important factor to consider is the radio's condition. Look for signs of wear and damage, such as scratches, dents, or missing parts. This will help you determine how much work will be required to restore the radio to its original condition.

By understanding the different types and models of vintage radios and knowing what to look for when identifying them, you'll be better equipped to build and enjoy your collection. With practice and patience, you'll soon become an expert in vintage radio identification.

4. FINDING AND ACQUIRING VINTAGE RADIOS:

Where to Look and What to Look For

Now that you have a good understanding of the different types and models of vintage radios, it's time to start building your collection. This chapter will provide you with tips and advice for finding and acquiring vintage radios.

One of the first places you should look for vintage radios is online. Websites such as eBay and Etsy offer a wide variety of vintage radios for sale, and you can often find good deals if you're patient and willing to shop around. However, it's important to be cautious when buying online, as you may not be able to inspect the radio in person before making a purchase. Look for sellers with good feedback ratings and a history of fair dealing, and ask plenty of questions before making a purchase.

Another good place to look for vintage radios is at antique shops and flea markets. These types of stores often have a wide selection of vintage radios, and you may be able to negotiate a better price if you're willing to haggle. When shopping in person, be sure to inspect the radio carefully before making a purchase. Look for signs of wear and damage, and ask the seller if the radio is in working condition.

You may also be able to find vintage radios at estate sales and auctions. These types of sales can be a great way to find rare and unusual radios, but you'll need to be prepared to compete with other collectors for the best finds. Make sure to arrive early and be prepared to bid aggressively if you find a radio you really want.

When looking for vintage radios, it's important to keep an eye out for certain features and characteristics. Look for radios that are in good condition, with minimal signs of wear or damage. It's also a good idea to look for radios that are complete, with all the original parts and components intact. And, of course, make sure to buy a radio that you find aesthetically pleasing and that fits with the overall theme and style of your collection.

By following these tips and taking the time to search for the perfect vintage radio, you'll soon be well on your way to building a collection that you can be proud of.

5. ASSESSING THE CONDITION OF VINTAGE RADIOS:

How to Evaluate Your Finds

As a vintage radio collector, one of the most important skills you'll need is the ability to assess the condition of the radios you find. Knowing how to evaluate the condition of a vintage radio will help you determine its value, as well as the amount of work required to restore it to its original condition.

When assessing the condition of a vintage radio, there are several factors to consider. The first is the overall appearance of the radio. Look for signs of wear and damage, such as scratches, dents, or missing parts. A radio in good condition will have minimal signs of wear and will be complete, with all original parts and components intact.

Next, evaluate the functionality of the radio. Does it turn on? Can you hear sound coming from the speakers? If the radio is not working, it may require some repairs or restoration work to get it functioning properly. However, keep in mind that some vintage radios may not be able to receive modern radio signals, so it's important to evaluate the functionality of the radio based on its original intended use.

Another important factor to consider is the originality of the radio. Look for signs of modification or repair, such as replaced parts or repainted cabinets. A radio that has been modified or restored may still be valuable, but it may not be as valuable as a radio that is completely original.

When evaluating the condition of a vintage radio, it's also important to consider the rarity and historical significance of the radio. A rare radio or one with a unique historical background may be more valuable, even if it's not in perfect condition.

Finally, consider the overall market demand for the type of radio you're evaluating. If there is high demand for a particular type of vintage radio, it may be worth more even if it's not in perfect condition.

By carefully assessing the condition of the vintage radios you find, you'll be able to make informed decisions about which radios to add to your collection and how much to pay for them. With practice and experience, you'll soon become an expert in evaluating the condition and value of vintage radios.

6. BASIC RADIO RESTORATION TECHNIQUES:

How to Clean, Repair and Restore Your

R estoring a vintage radio can be a rewarding and satisfying experience, but it can also be a complex and challenging task. In this chapter, we'll provide you with some basic radio restoration techniques to help you clean, repair, and restore your vintage radios.

Cleaning

The first step in restoring a vintage radio is to clean it. Over time, dust and dirt can accumulate inside the radio, which can cause interference and affect the sound quality. To clean your radio, you'll need to remove the back panel and use a soft brush or compressed air to remove any dust and debris. You can also use a gentle cleaning solution and a soft cloth to clean the outside of the radio.

Repair

Once your radio is clean, you can start to assess its condition and identify any areas that require repair. Common issues with vintage radios include broken tubes, damaged capacitors, and corroded wiring. To repair these issues, you'll need to have some basic knowledge of electronics and be comfortable using a soldering iron.

One of the most common repairs for vintage radios is to replace old or damaged capacitors. Capacitors store electrical energy and can degrade over time, which can cause distortion or a loss of sound quality. To replace a capacitor, you'll need to identify the faulty capacitor and carefully remove it using a soldering iron. You can then solder a new capacitor in its place.

Restoration

Once you've repaired any issues with your radio, you can start the process of restoring it to its original condition. This may involve repainting the cabinet, replacing missing parts, or polishing the knobs and dials. Depending on the extent of the restoration work, you may need to consult with a professional restoration expert or use specialized tools and equipment.

When restoring a vintage radio, it's important to be patient and take your time. Rushing the restoration process can result in damage to the radio or an unsatisfactory final result. With practice and experience, you'll soon become an expert in restoring vintage radios.

Conclusion

By using these basic radio restoration techniques, you'll be able to clean, repair, and restore your vintage radios to their original condition. Restoring a vintage radio can be a challenging and rewarding experience, and with the right tools and techniques, you can preserve these unique and historical pieces for future generations to enjoy.

7. ESSENTIAL TOOLS FOR RADIO COLLECTING:

A Guide to Must-Have Equipment

I f you're serious about radio collecting, it's essential to have the right tools to properly care for and maintain your vintage radios. In this chapter, we'll discuss the must-have equipment for radio collecting.

1. Multimeter

A multimeter is an essential tool for any radio collector. It allows you to measure voltage, resistance, and current, which is essential for troubleshooting and diagnosing issues with vintage radios.

2. Soldering Iron

A soldering iron is another must-have tool for radio collecting. It's essential for repairing and replacing components on vintage radios, such as capacitors, resistors, and wires.

3. Wire Cutters and Strippers

Wire cutters and strippers are useful for removing insulation from wires and cutting them to the right length. This is essential for repairing and replacing wires on vintage radios.

4. Screwdrivers

A set of screwdrivers in various sizes is important for removing the back panel and accessing the internal components of vintage radios. It's essential to have both flathead and Phillips head screwdrivers.

5. Pliers

Pliers are useful for bending wires and adjusting small components on vintage radios. They come in various sizes and types, such as needle-nose pliers, which are useful for working in tight spaces.

6. Deoxit Contact Cleaner

Deoxit contact cleaner is a must-have for cleaning the contacts and switches on vintage radios. It's a fast-acting cleaner that removes dirt, dust, and oxidation, which can cause interference and affect the sound quality.

7. Compressed Air Duster

A compressed air duster is useful for removing dust and debris from vintage radios, which can cause interference and affect the sound quality. It's also useful for cleaning hard-to-reach areas, such as inside the chassis.

8. Magnifying Glass

A magnifying glass is useful for inspecting small components and wires on vintage radios. It's essential for identifying and diagnosing issues with vintage radios.

9. Service Manual

A service manual is an essential resource for any radio collector. It provides detailed information on the components, wiring, and schematics of vintage radios, which is essential for troubleshooting and diagnosing issues.

Conclusion

By using these essential tools for radio collecting, you'll be able to properly care for and maintain your vintage radios. It's important to invest in high-quality tools and equipment to ensure that you're able to properly diagnose and repair any issues with your radios. With the right tools and techniques, you'll be able to preserve these unique and historical pieces for future generations to enjoy.

8. DISPLAYING YOUR RADIOS:

Creative Ways to Showcase
Your Collection

I f you're a radio collector, chances are you've amassed quite a collection of vintage radios. But what good is a collection if it's hidden away in a closet or attic? In this chapter, we'll discuss creative ways to showcase your collection and make it a focal point of your home.

1. Shelving Units

Shelving units are a classic way to display your radio collection. You can choose from a wide variety of styles, from sleek and modern to rustic and vintage. Make sure to arrange your radios in a way that showcases their unique designs and features.

2. Shadow Boxes

Shadow boxes are a great way to showcase individual radios or small groups of radios. You can mount the radios on a background of your choice, such as patterned paper or fabric, and add labels or descriptions to provide context.

3. Glass Cases

Glass cases are a more formal way to display your radio collection. You can choose from a variety of styles, from elegant antique display cases to sleek modern cabinets. Glass cases protect your radios from dust and damage while still allowing them to be admired.

4. Vintage Suitcases

Vintage suitcases can be repurposed as display cases for your radio collection. Simply remove the lid and add shelves to create a unique and eye-catching display.

5. Bookshelves

Bookshelves are a great option for smaller collections. You can mix your radios in with books and other decorative items to create a cohesive and interesting display.

6. Wall-Mounted Displays

Wall-mounted displays are a great way to showcase your radios without taking up valuable floor space. You can create a custom display by mounting individual radios on a decorative board or by arranging them in a creative pattern.

7. Pedestals

Pedestals are a great way to showcase individual radios or small groups of radios. You can choose from a variety of styles and materials, from sleek modern designs to rustic wooden pedestals.

8. Repurpose Old Radios

If you have old radios that are no longer functioning, consider repurposing them into unique display pieces. For example, you can turn an old radio into a lamp or use it as a decorative vase holder.

Conclusion

There are many creative ways to showcase your radio collection, from classic shelving units to unique repurposed displays. The key is to choose a display option that showcases the unique features and designs of your radios while also complementing your home decor. By displaying your collection, you'll be able to share your love of vintage radios with others and make your collection a focal point of your home.

9. CONNECTING YOUR RADIOS:

Tips for Playing and Listening to Your Vintage Radios

Once you've collected your vintage radios, you'll want to be able to play and listen to them. In this chapter, we'll provide tips and advice for connecting your radios and enjoying the vintage sounds they produce.

1. Safety First

Before connecting any vintage radio, make sure it's safe to use. If the wiring is frayed, the cord is damaged or the chassis is rusty, do not attempt to power up the radio. It's always best to have a qualified technician inspect and repair the radio before use.

2. Power Sources

Vintage radios typically require AC power, but some may also run on batteries. Make sure you have the appropriate power source for your radio, and that the voltage and current rating match the radio's requirements.

3. Antennas

Most vintage radios require an external antenna to receive signals. The antenna can be as simple as a length of wire attached to the radio's antenna input or a more elaborate outdoor antenna. Experiment with different antenna types to see which works best with your radio.

4. Grounding

Many vintage radios require a ground connection to function properly. This can be accomplished by attaching a wire to the radio's grounding terminal and connecting it to a water pipe or grounding rod. Make sure to follow proper electrical grounding procedures to avoid electric shock.

5. Speaker Connections

Vintage radios typically have an internal speaker, but some also have an external speaker output. If you're using an external speaker, make sure the speaker impedance matches the radio's output impedance to avoid damage to either component.

6. Tuning

Vintage radios typically have manual tuning controls, such as a dial or a knob. Take your time when tuning your radio to find the best reception, and be patient when making adjustments.

7. Maintenance

Regular maintenance is important to keep your vintage radios in top condition. This includes cleaning the radio's chassis and controls, replacing tubes and capacitors, and checking the power cord and antenna connections.

Conclusion

Connecting and playing vintage radios can be a rewarding experience, but it requires careful attention to safety and maintenance. Make sure to follow proper electrical procedures, experiment with different antennas and speakers, and maintain your radios regularly to keep them in top condition. With the right setup and care, you'll be able to enjoy the unique sounds of vintage radio for years to come.

10. THE ART OF RADIO CABINET REFINISHING:

How to Make Your Radios Look like New Again

One of the most enjoyable aspects of radio collecting is restoring vintage radios to their former glory. In this chapter, we'll discuss the art of radio cabinet refinishing, including the tools and techniques you'll need to make your radios look like new again.

1. Preparation

Before you start refinishing your radio cabinet, you'll need to remove any hardware and clean the cabinet thoroughly. This includes removing any old paint or varnish, filling in any cracks or holes, and sanding the cabinet to create a smooth surface.

2. Sanding

Once the cabinet is clean and dry, use a fine-grit sandpaper to sand the surface of the wood. This will remove any remaining finish and prepare the surface for the next steps.

3. Staining

If you want to change the color of the wood, now is the time to apply a stain. Choose a stain that matches the color you want to achieve and apply it evenly to the surface of the wood. Allow the stain to dry completely before moving on to the next step.

4. Varnishing

Once the stain has dried, apply a clear varnish to protect the wood and give it a shiny, new finish. Apply the varnish in thin, even coats, allowing each coat to dry completely before applying the next.

5. Reassembly

Once the varnish has dried, it's time to reattach any hardware and reassemble the radio. Be careful not to scratch the newly refinished surface as you reattach screws and knobs.

6. Maintenance

To keep your refinished radio looking like new, it's important to take proper care of it. This includes dusting the cabinet regularly and avoiding exposure to direct sunlight and extreme temperatures.

Conclusion

Refinishing a radio cabinet can be a rewarding and enjoyable process, but it requires careful attention to detail and patience. With the right tools and techniques, you can transform an old, worn-out radio into a beautiful piece of functional art.

11. RADIO COLLECTING ON A BUDGET:

Tips for Finding Affordable Radios

I f you're a beginner radio collector on a budget, finding affordable radios can be a great way to build your collection without breaking the bank. In this chapter, we'll explore some tips and tricks for finding affordable radios that are still in good condition.

1. Attend Swap Meets and Flea Markets

Swap meets and flea markets can be a great place to find affordable radios. These events often have vendors who specialize in vintage electronics, including radios. You can usually find a variety of models and price ranges at these events, making it easy to find something that fits your budget.

2. Check Online Marketplaces

Online marketplaces like eBay, Craigslist, and Facebook Marketplace can also be great resources for finding affordable radios. You can set up alerts for specific models or search for listings that are within your price range. Be sure to carefully read the descriptions and examine photos of the radio before making a purchase to ensure that it's in good condition.

3. Check Garage Sales and Estate Sales

Garage sales and estate sales can be another great place to find affordable radios. These events often have items for sale that are priced to move, making it easier to find a bargain. If you're looking for a specific type of radio, consider reaching out to the seller in advance to see if they have any radios that match your interests.

4. Look for Non-Working Radios

If you're handy with electronics, you may be able to find a non-working radio for a bargain price and then repair it yourself. This can be a great way to save money and learn more about the inner workings of radios. However, be sure to carefully assess the extent of the repairs that will be needed before making a purchase.

5. Join a Radio Collecting Club

Joining a radio collecting club can be a great way to meet other collectors and learn more about the hobby. These clubs often have members who are willing to sell or trade radios, which can be a great way to find affordable models. Additionally, club members may be able to offer advice on where to find affordable radios in your area.

In conclusion, finding affordable radios for your collection doesn't have to be difficult. By attending swap meets and flea markets, checking online marketplaces, looking for non-working radios, and joining a radio collecting club, you can find great deals on vintage radios that fit your budget.

12. UNDERSTANDING RADIO ELECTRONICS:

A Beginner's Guide to Basic Circuitry and Components

Radio electronics can seem intimidating at first, but with a little bit of knowledge about basic circuitry and components, anyone can get started in this fascinating hobby. In this beginner's guide, we'll introduce you to the basics of radio electronics.

1. Basic Circuitry

At its most basic level, a radio circuit consists of three components: a source of power, a device that can detect radio waves (like an antenna), and a device that can amplify and tune the signal (like a radio tuner). When a radio wave is detected by the antenna, it generates a tiny electrical current that can be amplified and tuned by the radio tuner, allowing you to listen to the signal.

2. Resistors

A resistor is a component that limits the flow of electrical current in a circuit. They are typically made of a material with high resistance, such as carbon or metal. Resistors are often used in radio circuits to control the amount of current flowing through a specific part of the circuit.

3. Capacitors

A capacitor is a component that can store electrical charge. They are typically made of two conductive plates separated by a non-conductive material. Capacitors are often used in radio circuits to store energy and release it when needed, smoothing out fluctuations in the electrical current.

4. Inductors

An inductor is a component that can store energy in a magnetic field. They are typically made of a coil of wire wrapped around a core made of magnetic material. Inductors are often used in radio circuits to tune a specific frequency or filter out unwanted frequencies.

5. Diodes

A diode is a component that allows electrical current to flow in only one direction. They are typically made of a semiconductor material such as silicon or germanium. Diodes are often used in radio circuits to convert AC (alternating current) to DC (direct current) or to demodulate (extract the audio signal from) an AM radio signal.

6. Transistors

A transistor is a component that can amplify or switch electrical signals. They are typically made of semiconductor material and consist of three layers: the emitter, the base, and the collector. Transistors are often used in radio circuits to amplify weak signals or switch between different parts of the circuit.

7. Integrated Circuits

An integrated circuit (IC) is a complete electronic circuit that is contained within a single chip. They are often used in modern radios to perform functions like tuning, amplification, and audio processing. ICs can be more efficient and reliable than discrete components, but they can also be more difficult to troubleshoot and repair.

In conclusion, understanding basic circuitry and components

is essential for anyone interested in radio electronics. By familiarizing yourself with resistors, capacitors, inductors, diodes, transistors, and integrated circuits, you can begin to explore the fascinating world of radio electronics and build your own circuits and devices.

13. COLLECTING AND RESTORING ANTIQUE RADIO TUBES:

The Heart of Vintage Radios

A ntique radio tubes, also known as vacuum tubes or valves, are an essential component of vintage radios. These tubes were used in radios from the early 1900s through the 1960s and played a crucial role in the development of radio technology. In this article, we'll explore the process of collecting and restoring antique radio tubes.

1. Understanding Vacuum Tubes

A vacuum tube is an electronic device that consists of a glass or metal envelope containing electrodes and a vacuum. When electrical current is applied to the electrodes, it can be amplified and modulated to create an audio signal. Vacuum tubes were used in early radios, televisions, and other electronic devices before being replaced by transistors and integrated circuits.

2. Collecting Antique Radio Tubes

Collecting antique radio tubes can be a rewarding hobby, but it requires some knowledge and research. To get started, you'll need to identify the types of tubes that were used in specific radio models. You can use online resources or reference books to find this information.

Once you know what types of tubes you're looking for, you can begin searching for them at antique stores, online marketplaces, and radio swap meets. Be sure to carefully inspect the tubes for damage or signs of wear, as tubes that are in poor condition may not function properly.

3. Restoring Antique Radio Tubes

Restoring antique radio tubes requires some specialized knowledge and equipment. If you're new to tube restoration, it's best to start with simple repairs and work your way up to more complex projects.

To restore a tube, you'll need to remove it from the radio and clean it using specialized solvents and brushes. You'll then need to test the tube using a tube tester to determine whether it's functioning properly. If the tube is not working correctly, you may need to replace one or more components inside the tube, such as the filament or grid.

4. Caring for Antique Radio Tubes

To keep your antique radio tubes in good condition, it's important to store them properly. Tubes should be stored in a dry, cool location away from direct sunlight. Avoid exposing them to temperature extremes or moisture, as this can damage the sensitive internal components.

In conclusion, collecting and restoring antique radio tubes can be a rewarding hobby for anyone interested in vintage radios. By understanding vacuum tubes, identifying the types of tubes used in specific radio models, and learning how to restore and care for tubes, you can build a collection of functioning tubes and keep your vintage radios working for years to come.

14. ADVANCED RESTORATION TECHNIQUES:

Repairing and Replacing Components

A dvanced restoration techniques for antique radios involve repairing and replacing components that may be damaged or no longer functioning properly. These techniques require a deeper understanding of the internal workings of vintage radios and specialized equipment to perform repairs safely and effectively. In this article, we'll explore the process of repairing and replacing components in antique radios.

1. Diagnosing Problems

Before beginning any repair or replacement work, it's essential to diagnose the problem with the radio. This may involve testing individual components, checking connections, and troubleshooting the overall circuitry of the radio. Once you've identified the source of the problem, you can determine which components need repair or replacement.

2. Repairing Components

In some cases, damaged components can be repaired rather than replaced. This may involve re-soldering loose connections, replacing damaged wires, or repairing cracks in the radio's chassis. Careful attention to detail and a steady hand are required to perform these repairs, as the delicate components inside antique radios can be easily damaged.

3. Replacing Components

If a component is beyond repair, it will need to be replaced. This may involve finding a replacement part from a salvage yard or online marketplace, or sourcing a new component from a specialized supplier. Replacement components may include resistors, capacitors, inductors, transformers, and vacuum tubes.

When replacing components, it's essential to ensure that the new component is the correct size and type for the radio. Replacing a component with one that has a different value or type can cause the radio to malfunction or even cause further damage to other components.

4. Calibration and Alignment

After replacing components, it's essential to calibrate and align the radio to ensure that it's working correctly. This may involve adjusting the radio's frequency, sensitivity, and selectivity to optimize its performance. Specialized equipment such as an oscilloscope, signal generator, and wattmeter may be required to perform these adjustments.

In conclusion, advanced restoration techniques for antique radios involve diagnosing problems, repairing and replacing components, and calibrating and aligning the radio for optimal performance. These techniques require specialized knowledge and equipment, but with practice, anyone can learn to restore vintage radios to their former glory.

15. THE FUTURE OF RADIO COLLECTING:

How to Connect with Other Collectors and Find Resources

As technology continues to evolve, the future of radio collecting is becoming increasingly exciting. With advancements in digital technology, vintage radios can now be restored and even enhanced with modern features. In this article, we'll explore how to connect with other collectors and find resources to stay up to date on the latest developments in the world of radio collecting.

1. Joining Clubs and Groups

One of the best ways to connect with other radio collectors is to join a club or group. There are many organizations that specialize in vintage radios, such as the Antique Wireless Association and the Antique Radio Club of America. These clubs offer resources, events, and opportunities to connect with other collectors and enthusiasts.

Online forums and social media groups are also great resources for connecting with other collectors. Facebook groups, Reddit threads, and online forums like RadioReference.com and Radio Museum offer a space to ask questions, share resources, and connect with other collectors.

2. Attending Events and Swap Meets

Attending events and swap meets is another great way to connect with other collectors and find resources. These events offer an opportunity to see and buy vintage radios, as well as to connect with other collectors and enthusiasts. Events like the Winterfest Antique Radio Auction and the Midwest Antique Radio Club's annual meet are just a few examples of the many events and swap meets that take place throughout the year.

3. Using Online Resources

There are many online resources available for radio collectors. Websites like Antique Radios and Radio Attic offer a wealth of information on vintage radios, including model information, repair guides, and restoration tips. RadioWorld and Radio Magazine are also great resources for staying up to date on the latest developments in the world of radio collecting.

4. Learning New Skills

As the technology used in vintage radios becomes more complex, it's important for collectors to stay up to date on the latest repair and restoration techniques. There are many resources available for learning new skills, including online tutorials, books, and courses. The ARRL (American Radio Relay League) offers courses in radio repair and restoration, and the National Association of Radio and Telecommunications Engineers offers a certification program in radio repair.

In conclusion, the future of radio collecting is bright, and there are many resources available for connecting with other collectors and staying up to date on the latest developments in the field. By joining clubs and groups, attending events and swap meets, using online resources, and learning new skills, anyone can become a successful and knowledgeable radio collector.